STARTING OUT

GARBAGE TRUCKS

MEG GREVE

CREATIVE EDUCATION • CREATIVE PAPERBACKS

CONT

ENTS

I SEE A GARBAGE TRUCK.

An arm reaches out and grabs the can. It looks like a robot!

This truck is a side loader.

Trash gets squeezed and smashed.

The garbage goes into a hopper.
It gets pushed by a blade.

This garbage truck is
a front loader.
Up goes the dumpster.
In goes the garbage.

Workers move the cans. They keep the neighborhood clean.

Thank you! See you next week!

MAKE A NOISE

RRRRRR!

Can you make the sounds of a garbage truck?

Listen to these sounds:

https://www.youtube.com/watch?v=PbO9rCAhswo

Now it is your turn!

GARBAGE TRUCK WORDS

blade: a metal part on a garbage truck that smashes the trash to make more room

dumpster: a large metal container that holds lots of trash

hopper: a place in the garbage truck where the trash is kept

loader: the part of a garbage truck that puts the trash in the hopper

READING CORNER

Dickmann, Nancy. *Garbage Trucks*. Mankato, Minn.: Capstone, 2022.

Kaiser, Brianna. *All About Garbage Collectors*. Minneapolis, Minn.: Lerner, 2022.

Keane, Max. *Trash Truck*. New York: HarperCollins, 2022.

INDEX

PUBLISHED BY CREATIVE EDUCATION AND CREATIVE PAPERBACKS
P.O. Box 227, Mankato, Minnesota 56002
Creative Education and Creative Paperbacks are imprints of The Creative Company
www.thecreativecompany.us

LIBRARY OF CONGRESS CATALOGING-IN-PUBLICATION DATA
Names: Greve, Meg, author.
Title: Garbage trucks / by Meg Greve.
Description: Mankato, Minnesota : Creative Education and Creative Paperbacks, [2025] | Series: Starting out | Includes bibliographical references and index. | Audience: Ages 4-7 | Audience: Grades K-1 | Summary: "Garbage trucks will introduce budding book learners to a noisy, colorful world with this new Starting Out title. Colorful photos, labeled diagrams, 'Make a Noise' section, glossary, and more ignite a passion for learning"-- Provided by publisher.
Identifiers: LCCN 2023059448 (print) | LCCN 2023059449 (ebook) | ISBN 9798889891741 (library binding) | ISBN 9781682775592 (paperback) | ISBN 9798889891864 (ebook)
Subjects: LCSH: Refuse collection vehicles--Juvenile literature. | Refuse collection--Juvenile literature. | CYAC: Refuse collection vehicles. | LCGFT: Instructional and educational works.
Classification: LCC TD794 .G74 2025 (print) | LCC TD794 (ebook) | DDC 628.4/420284--dc23/eng/20240128
LC record available at https://lccn.loc.gov/2023059448
LC ebook record available at https://lccn.loc.gov/2023059449

DESIGN AND PRODUCTION
Design by Rhea Magaro and Jennifer Bowers
Production by Beeline Media and Design
Art direction by Tom Morgan
Printed in the United States of America

PHOTOGRAPHS by Getty Images (ansonsaw), Shutterstock (I'm friday, freestore 839, Aastels, Another77, Noel V. Baebler, Mirror-Images, degetzica, brajianni, Victor Yarmolyuk, Sirisak_baokaew, Ljupco Smokovski, Gargantiopa, Bohbeh)